MANDALA
COLORING BOOK FOR STRESS REFIEF

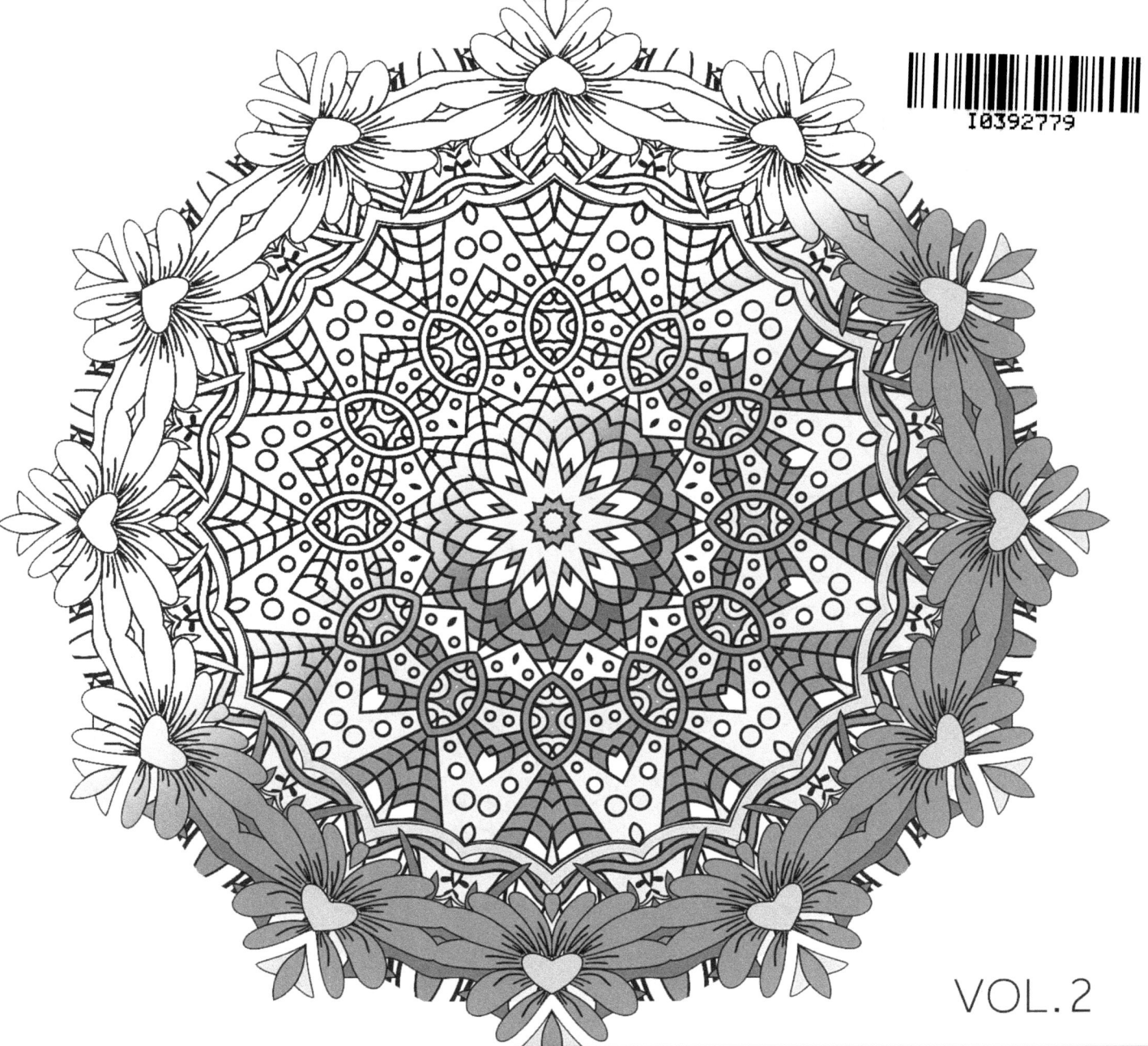

VOL. 2

ADULT COLORING BOOK DESIGNS

COLOR TEST PAGE

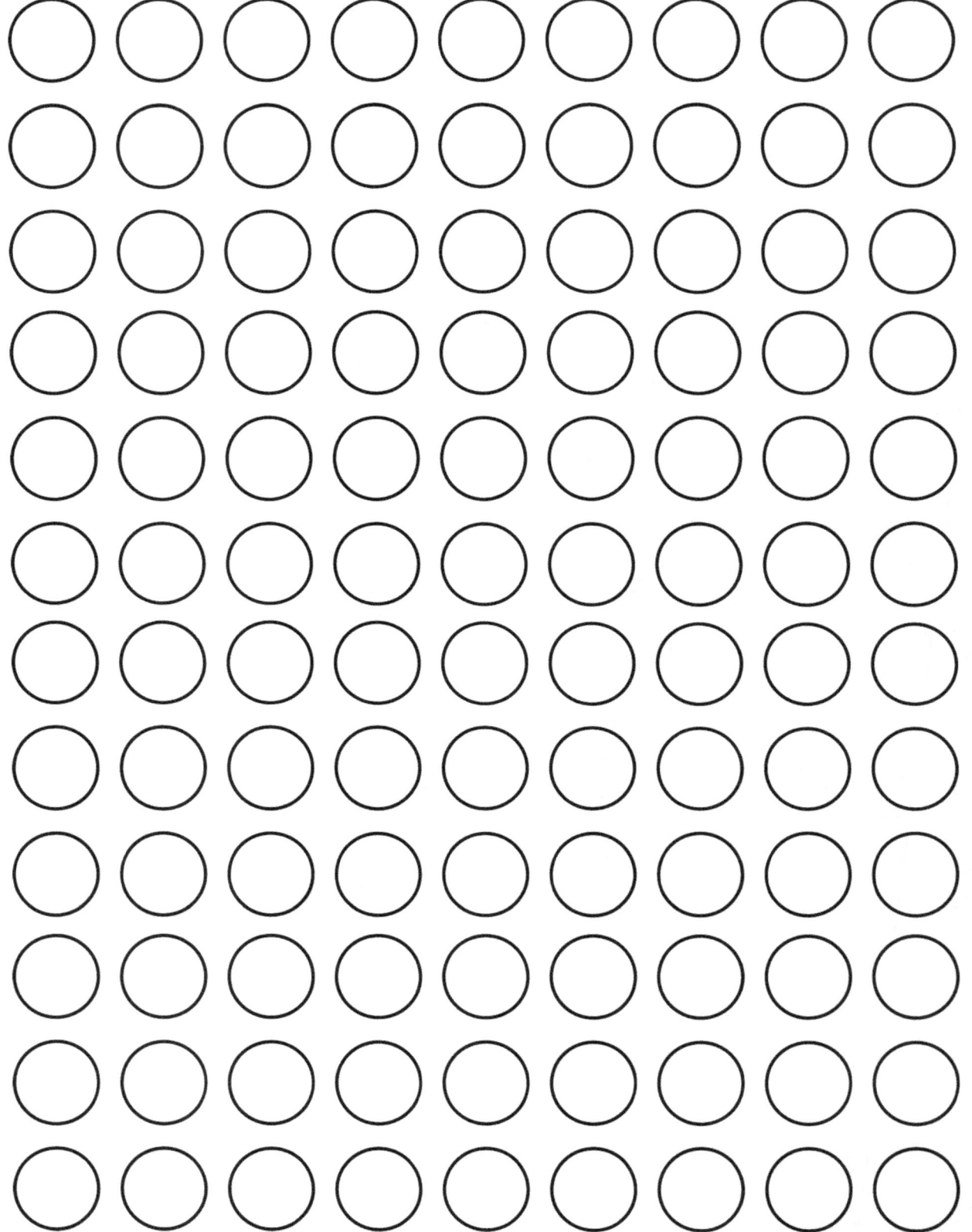

COLOR TEST PAGE

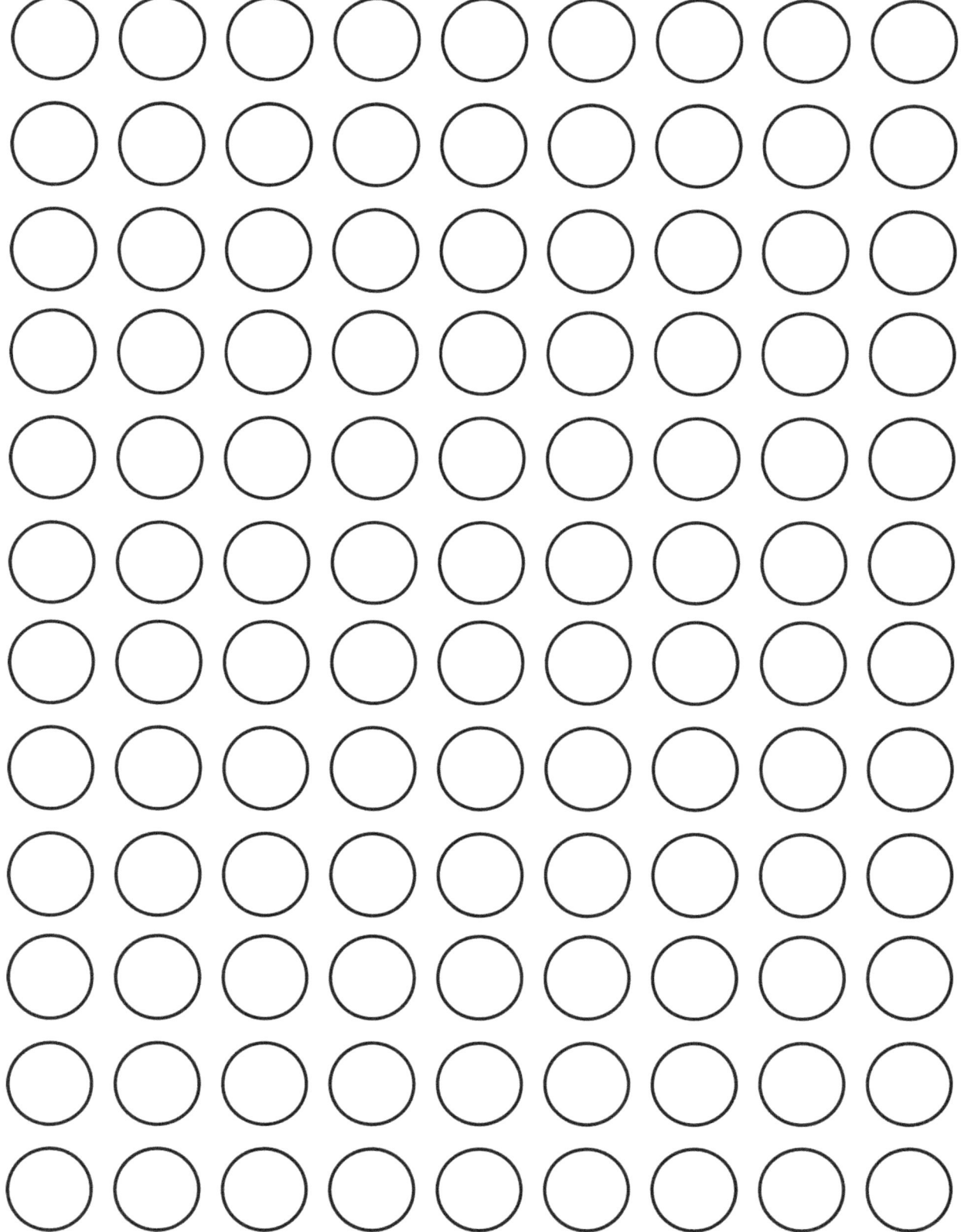

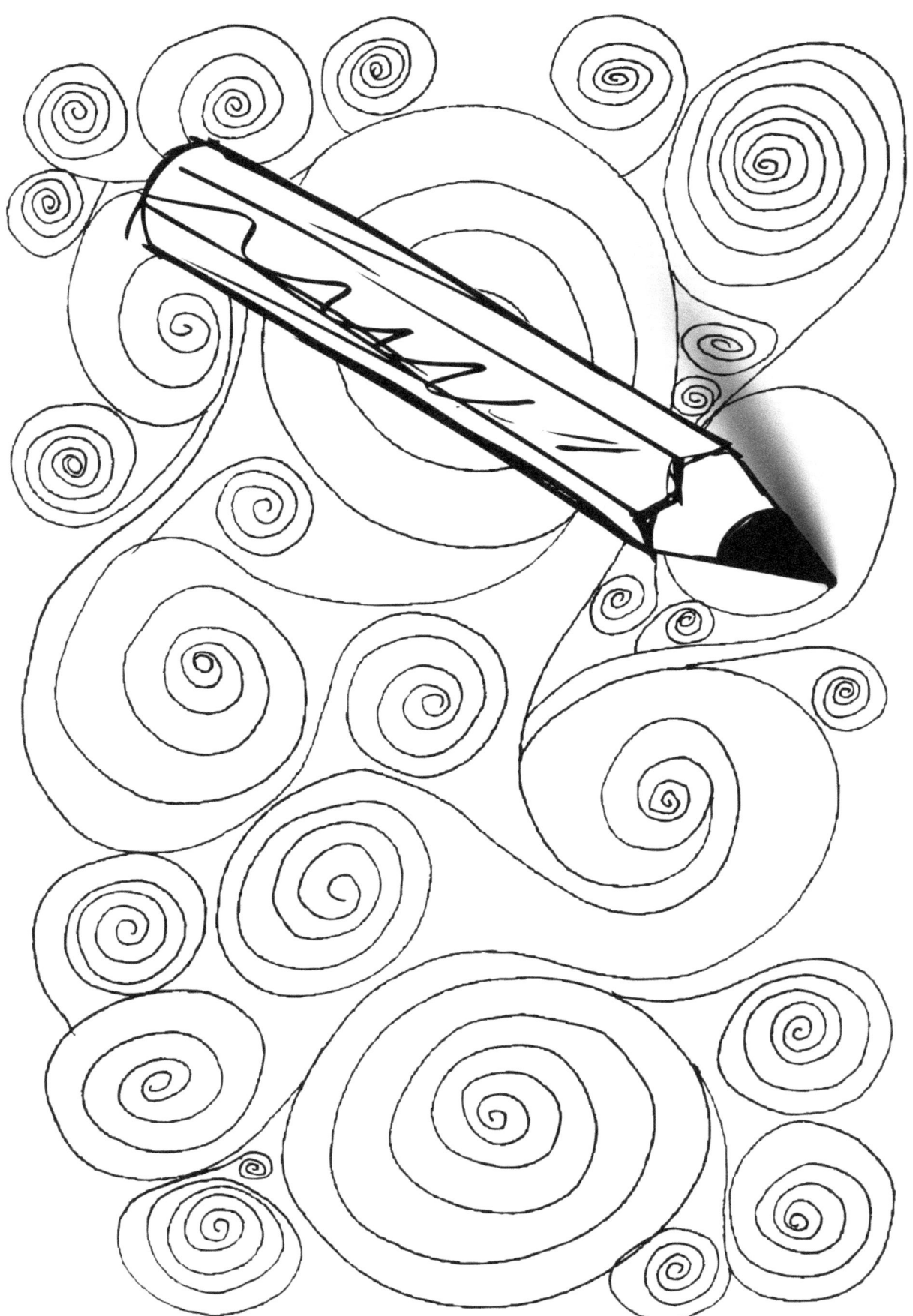

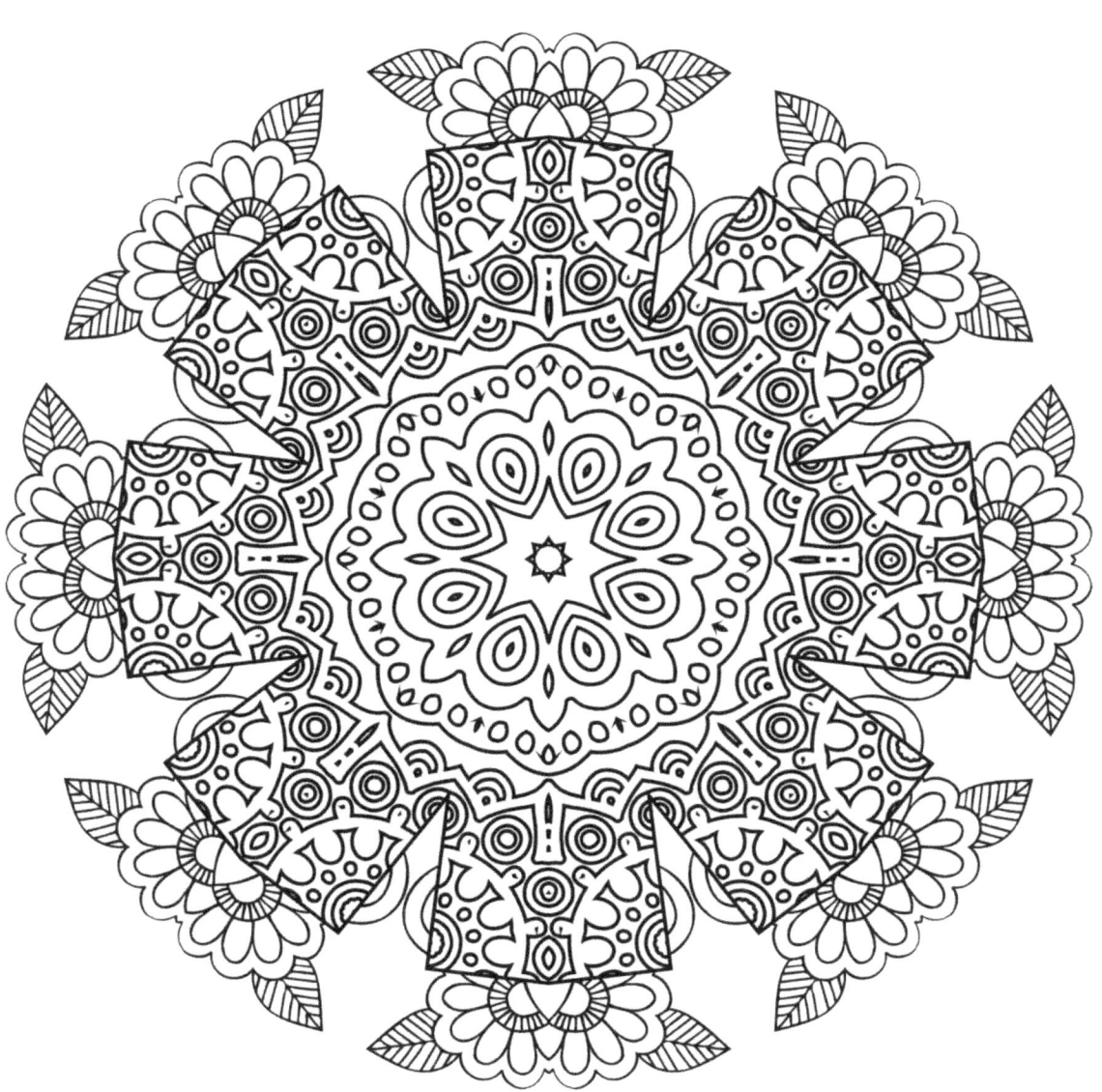

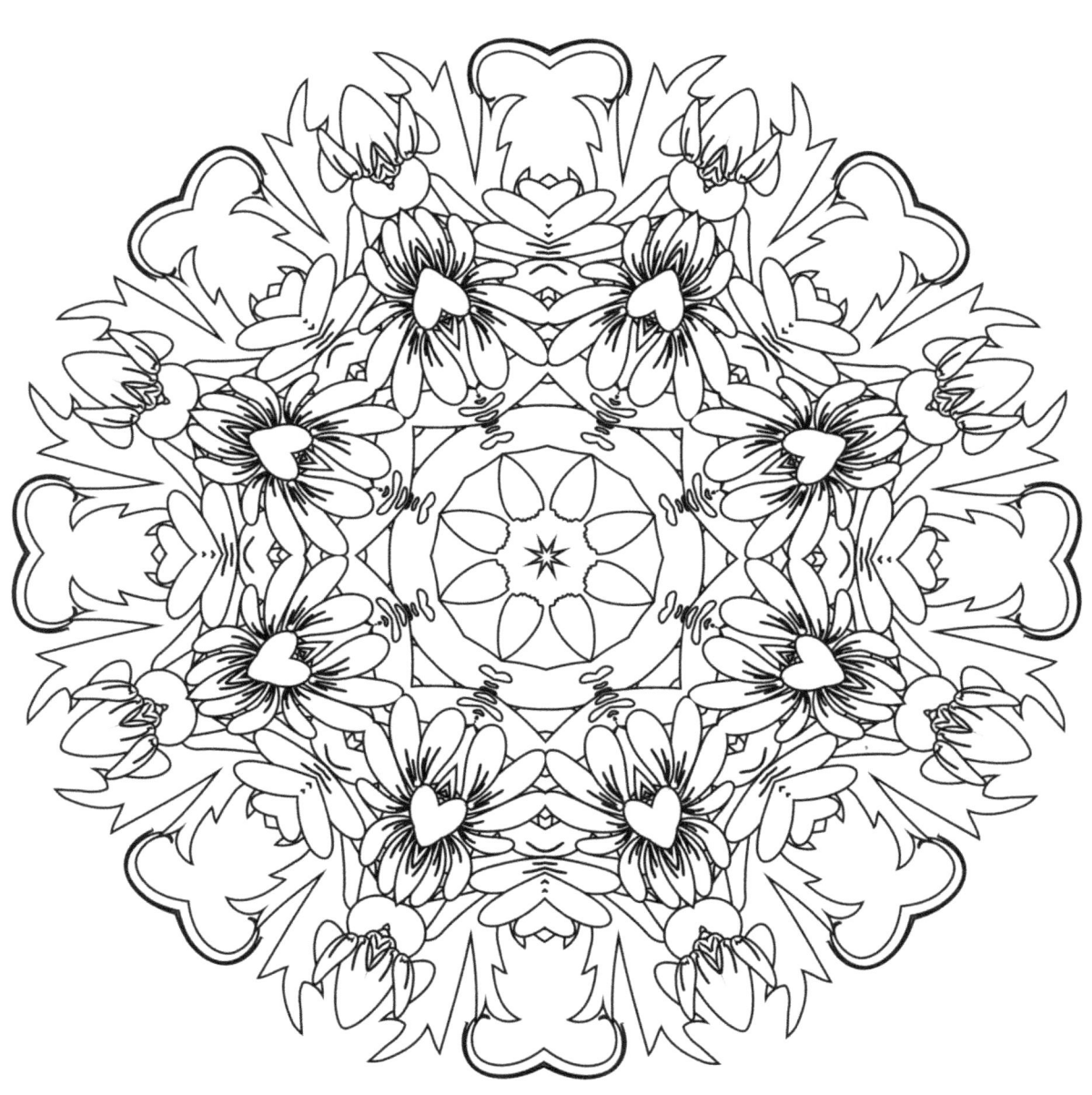

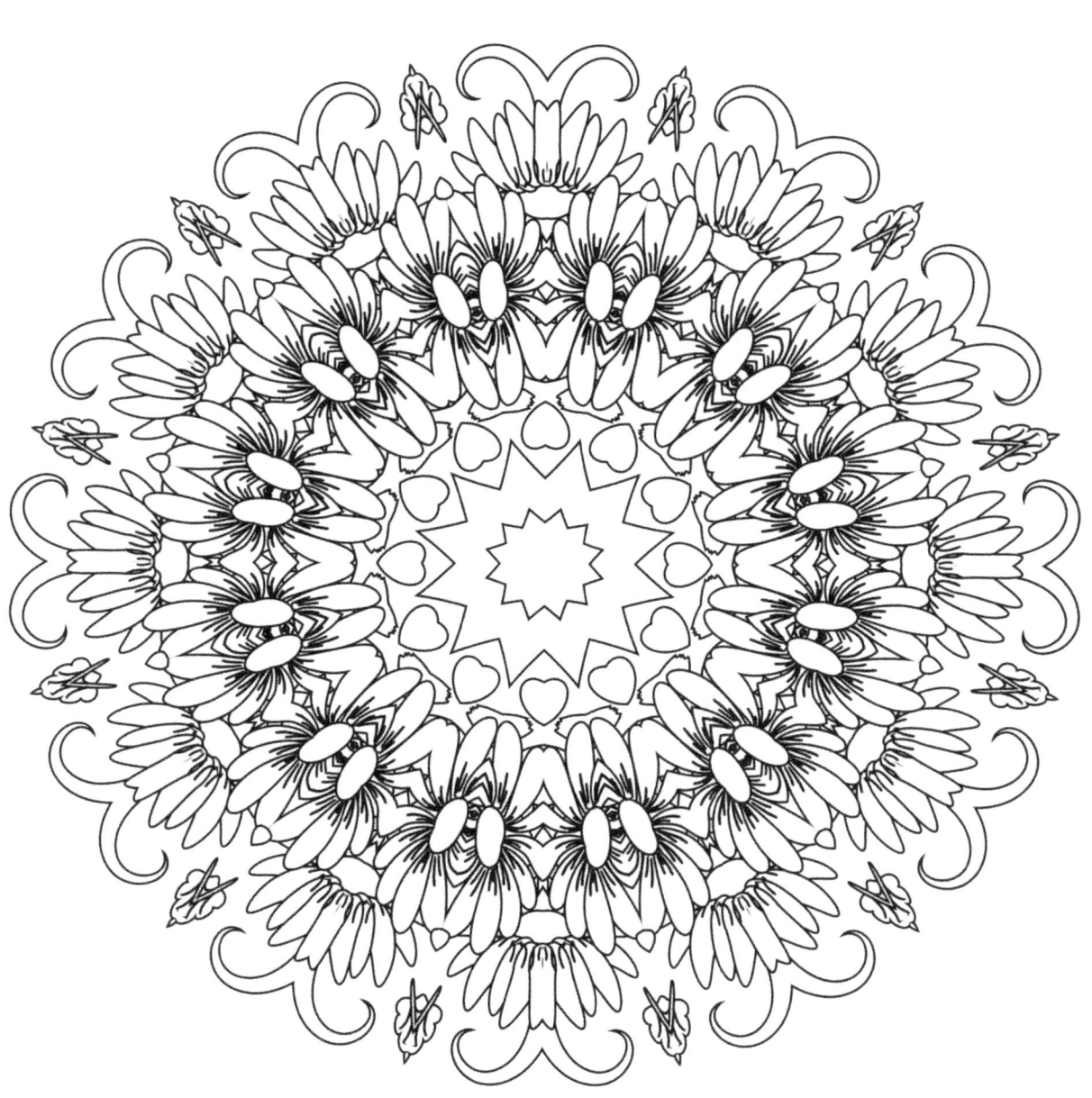

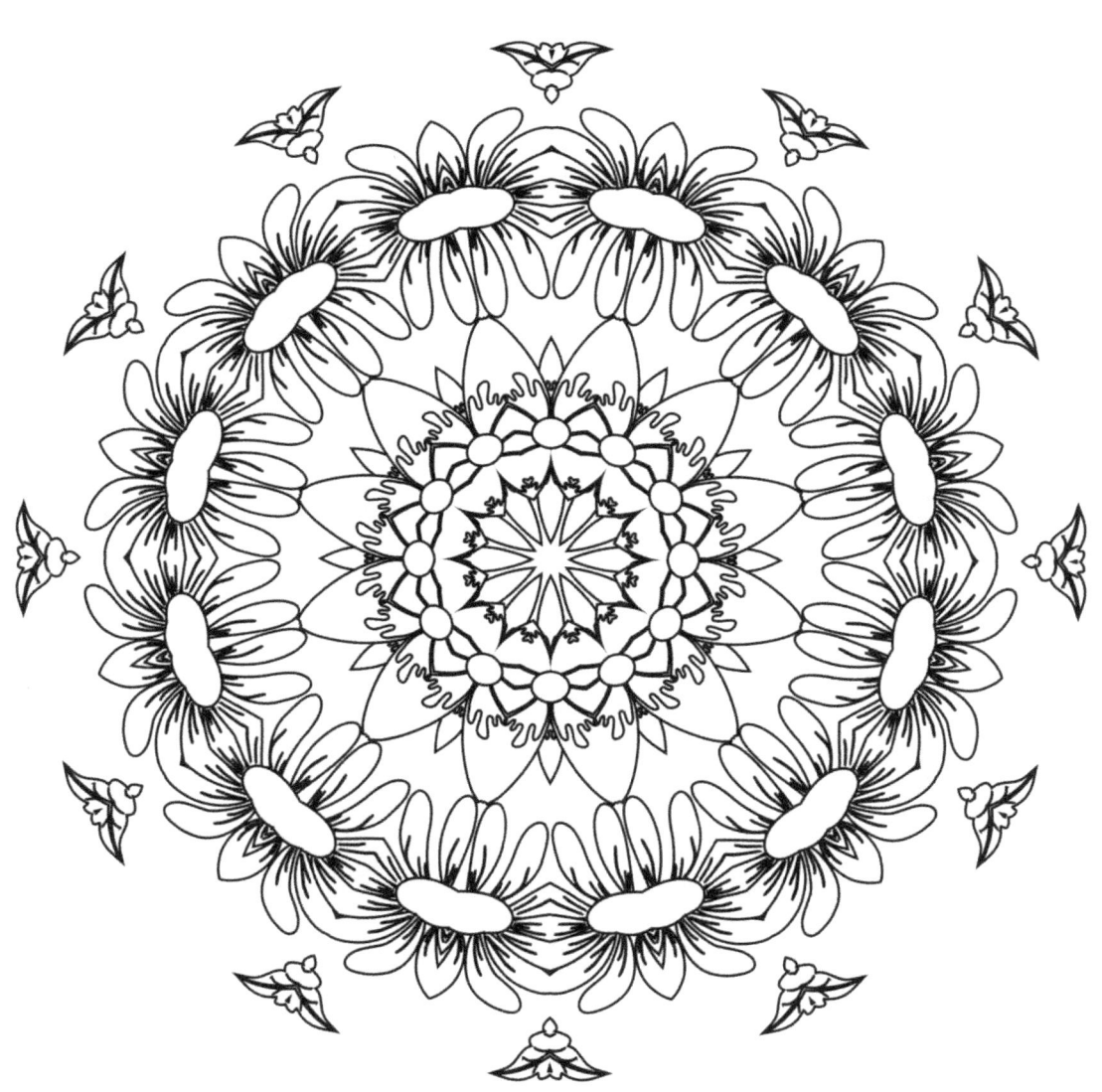

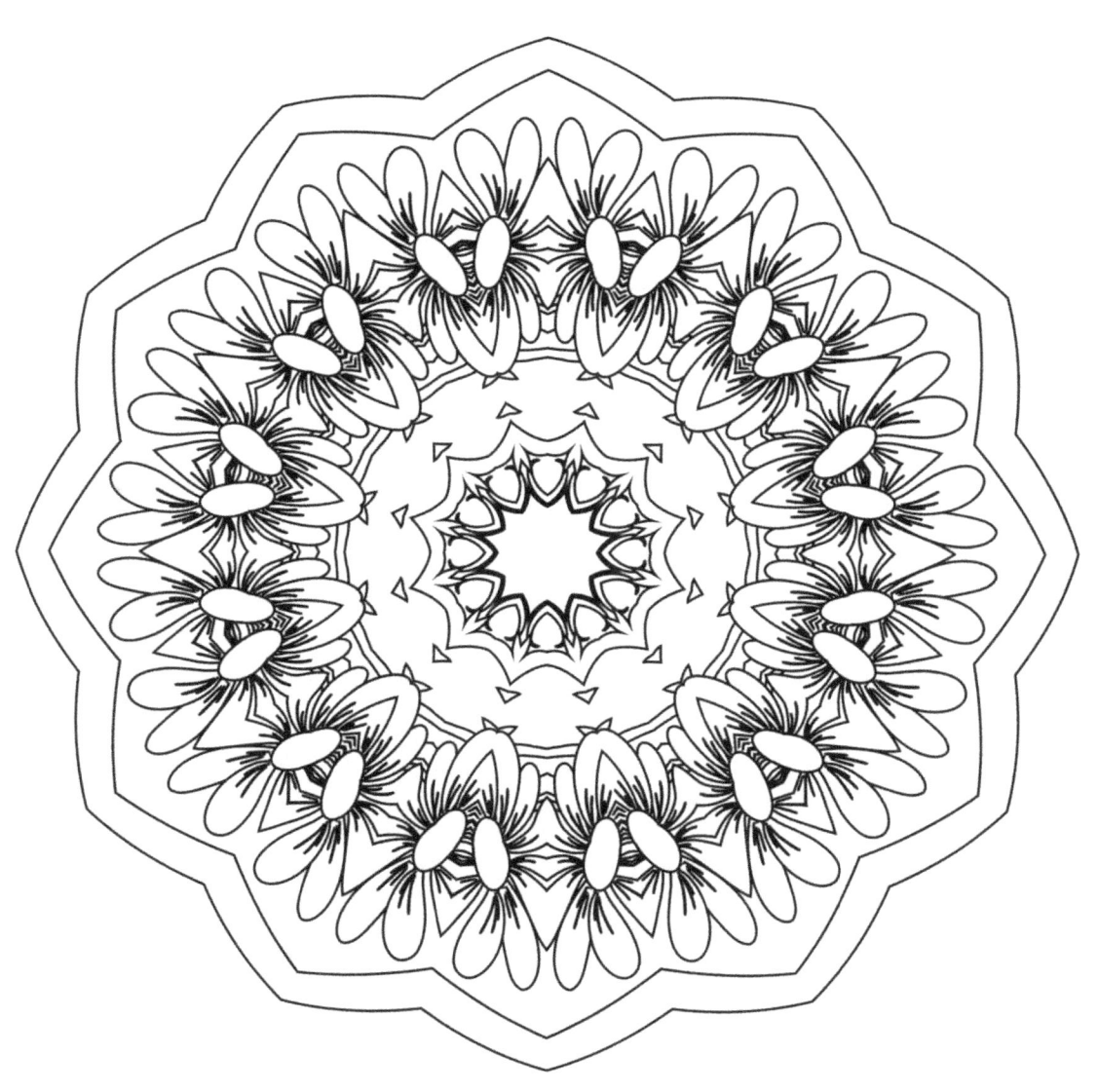

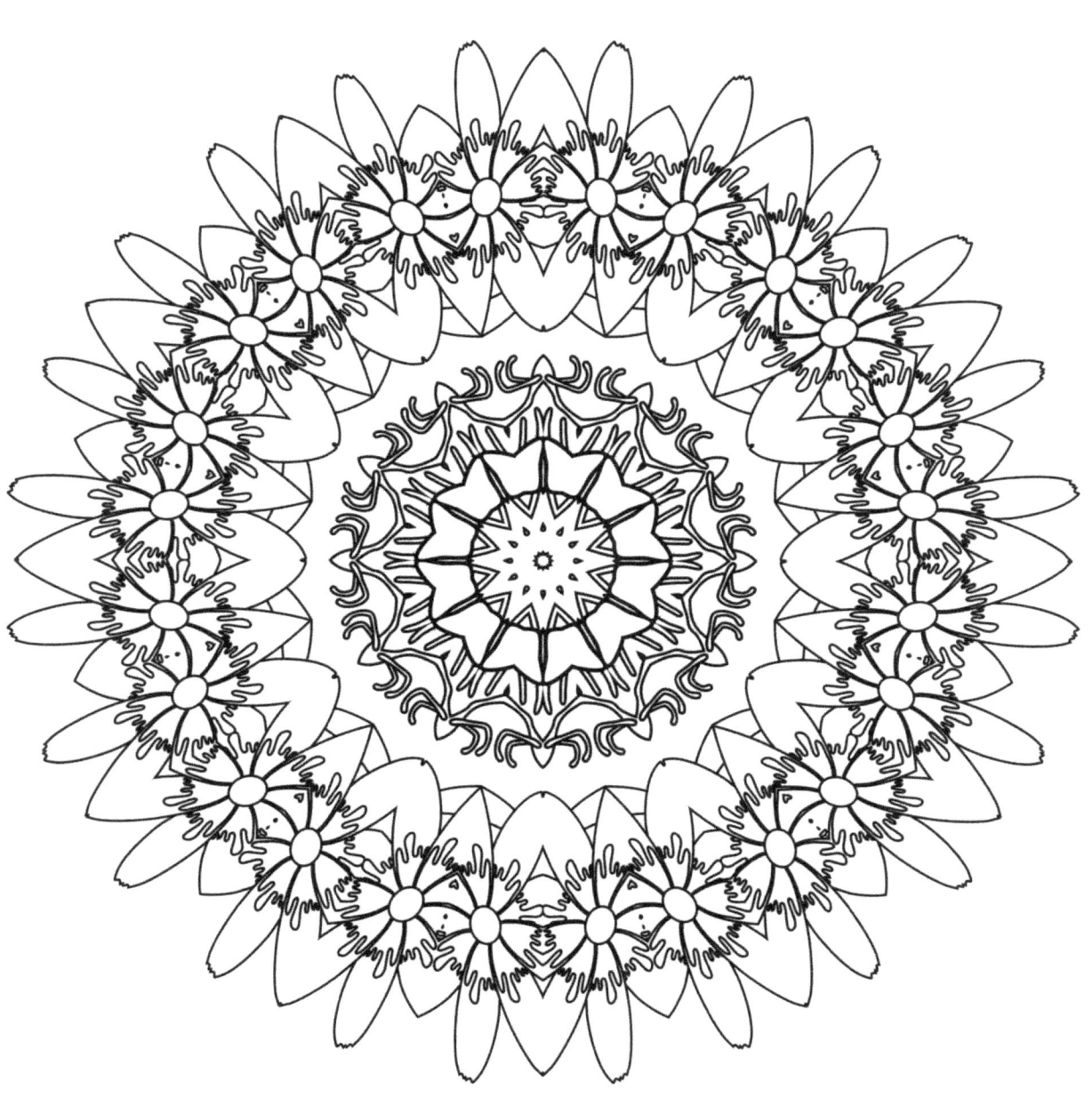

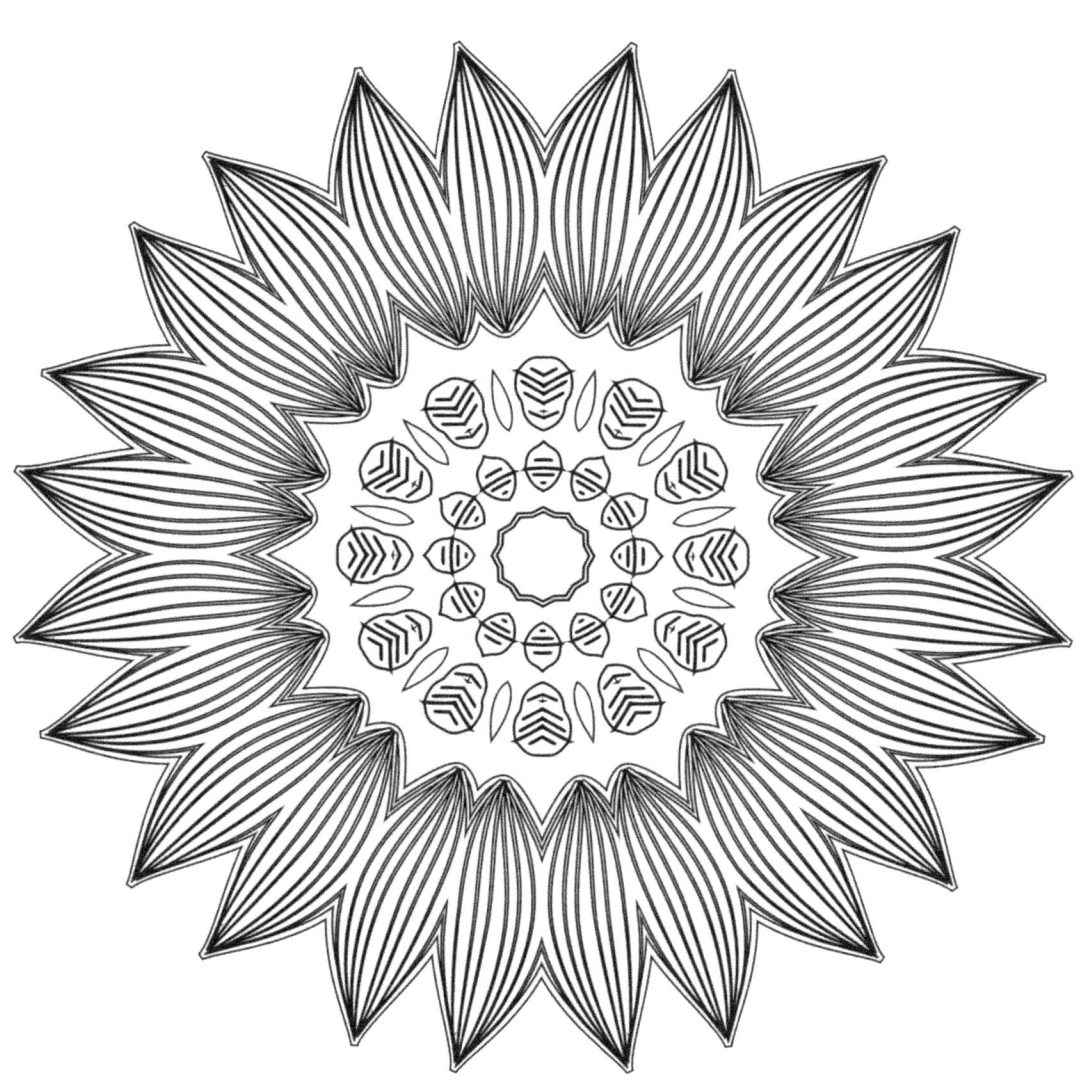

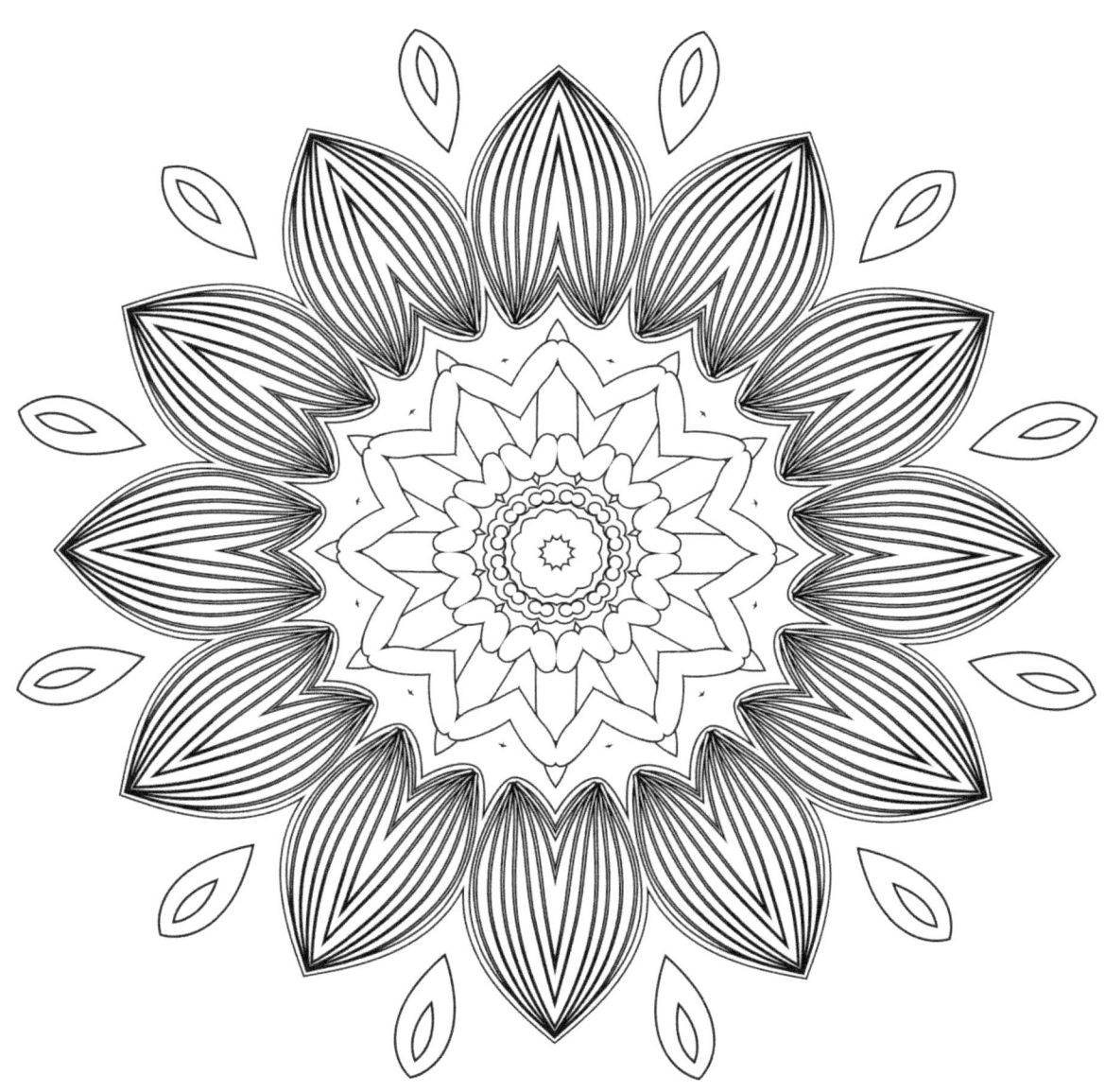

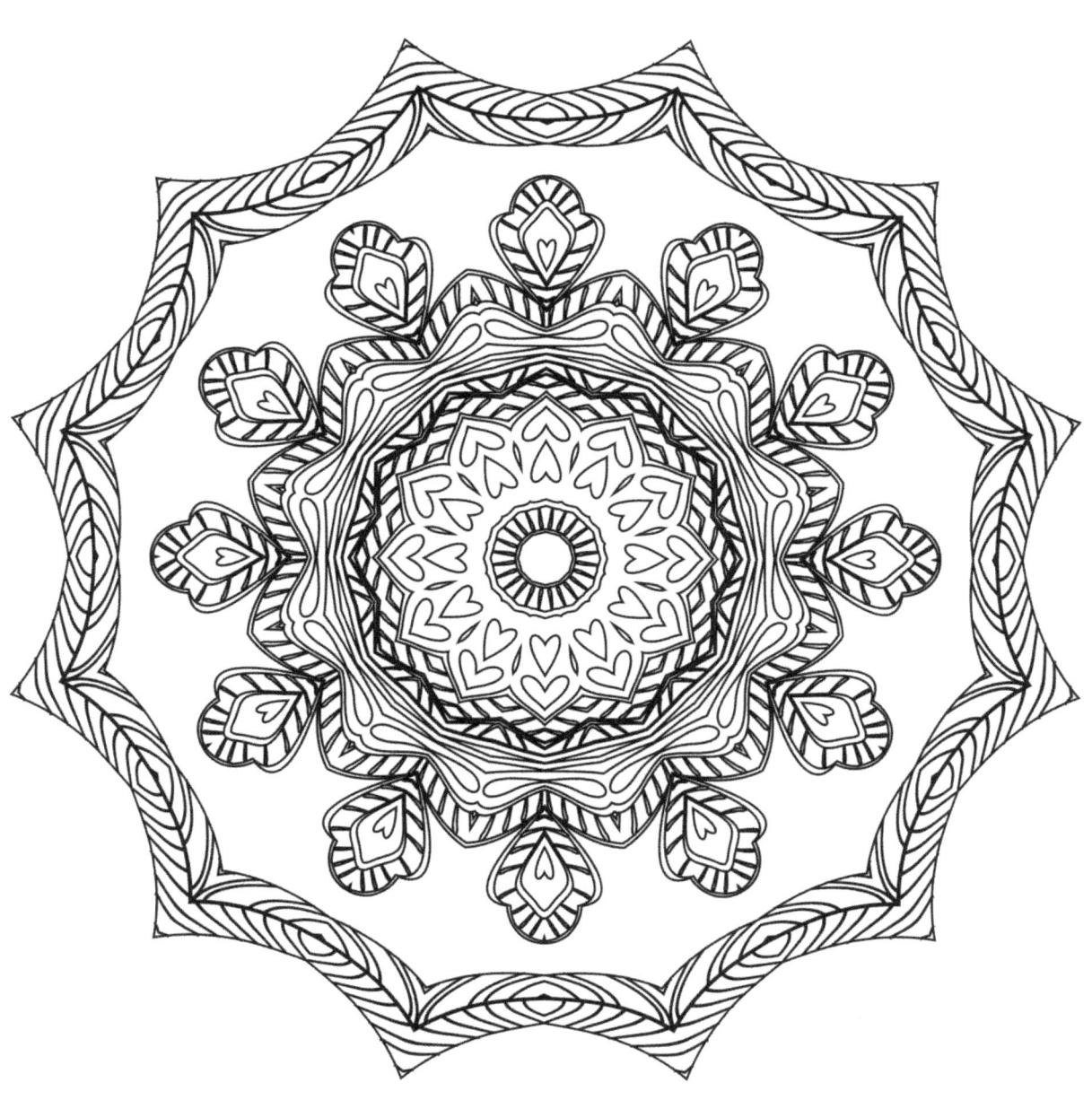

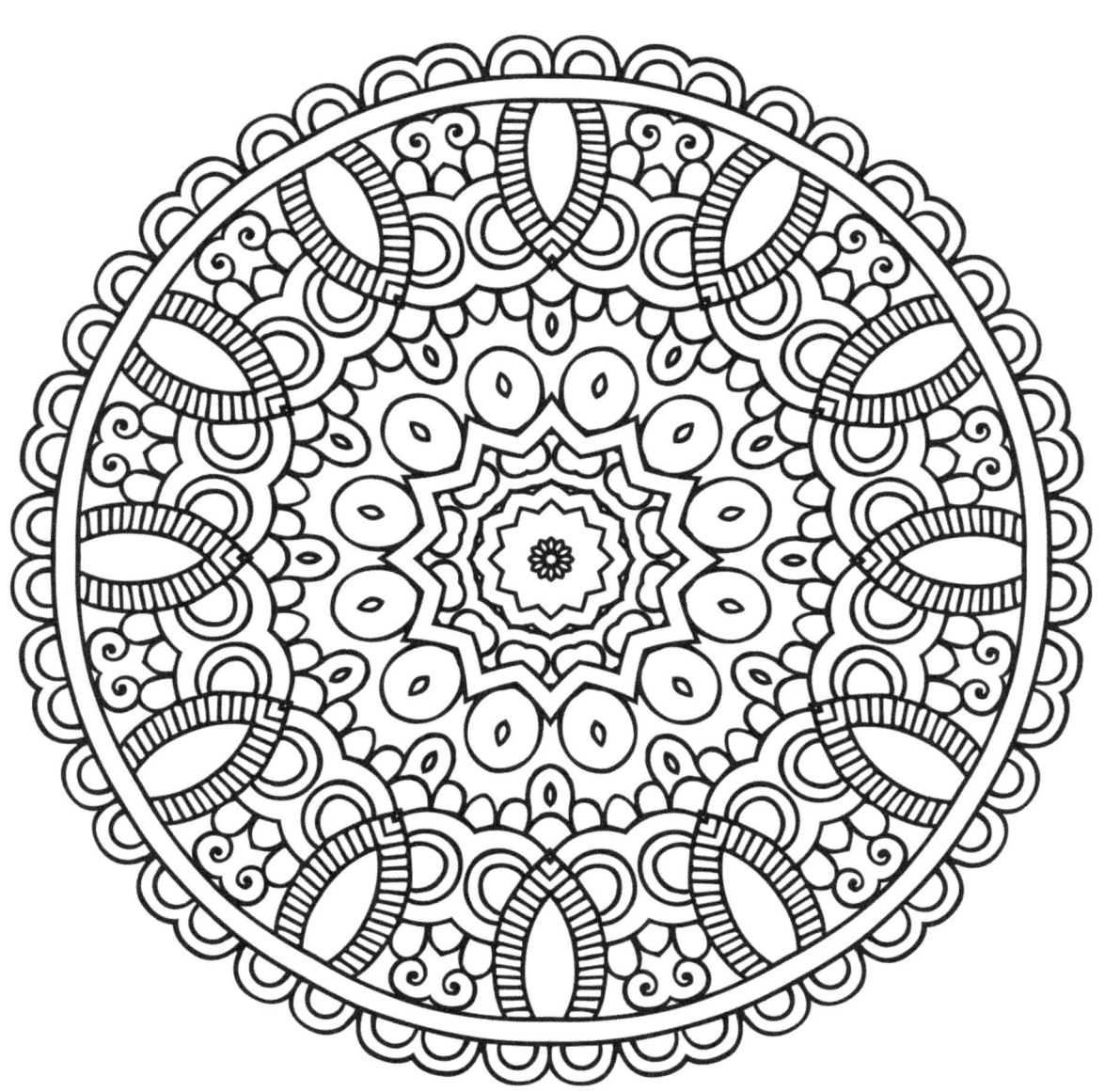

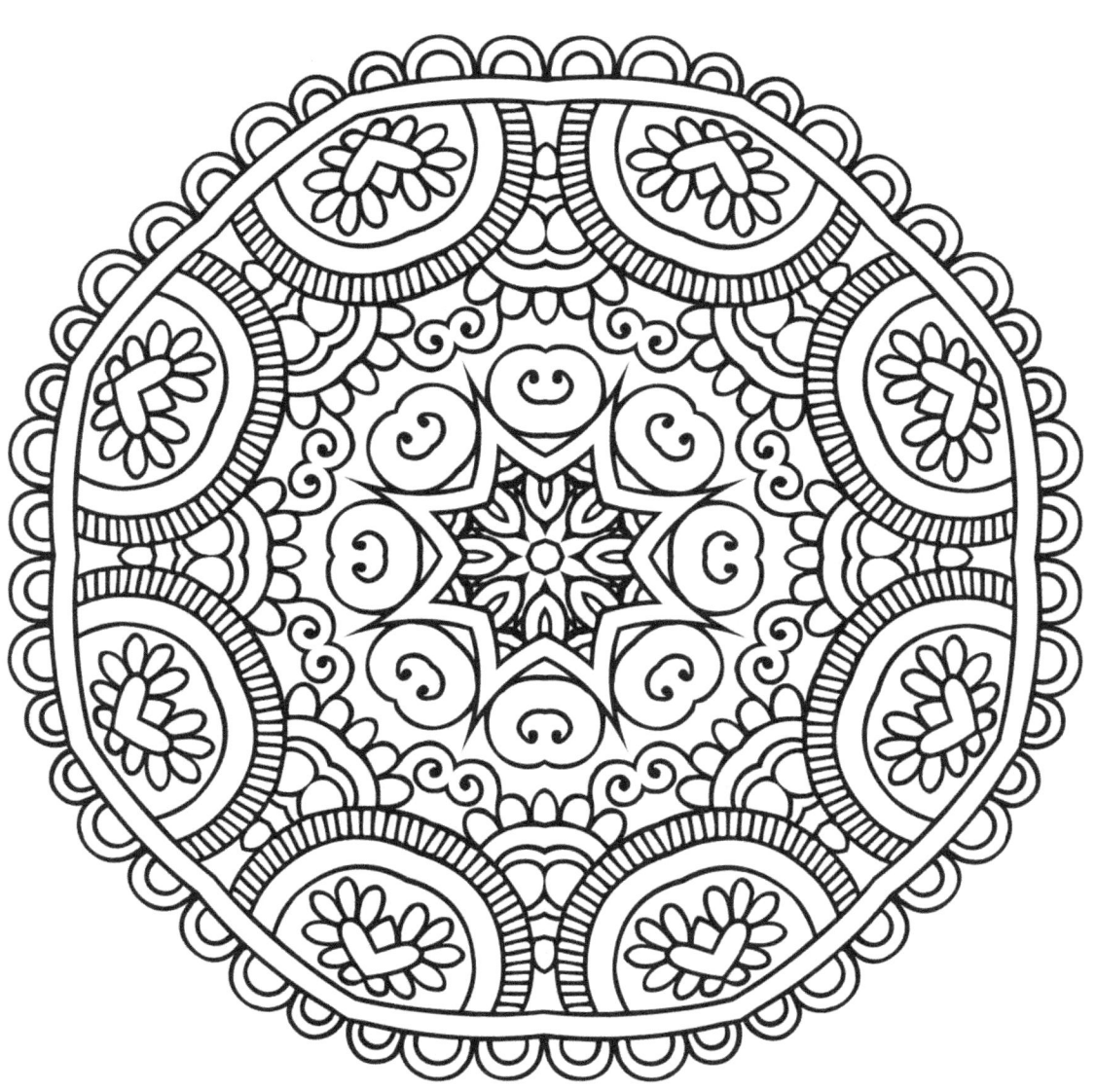

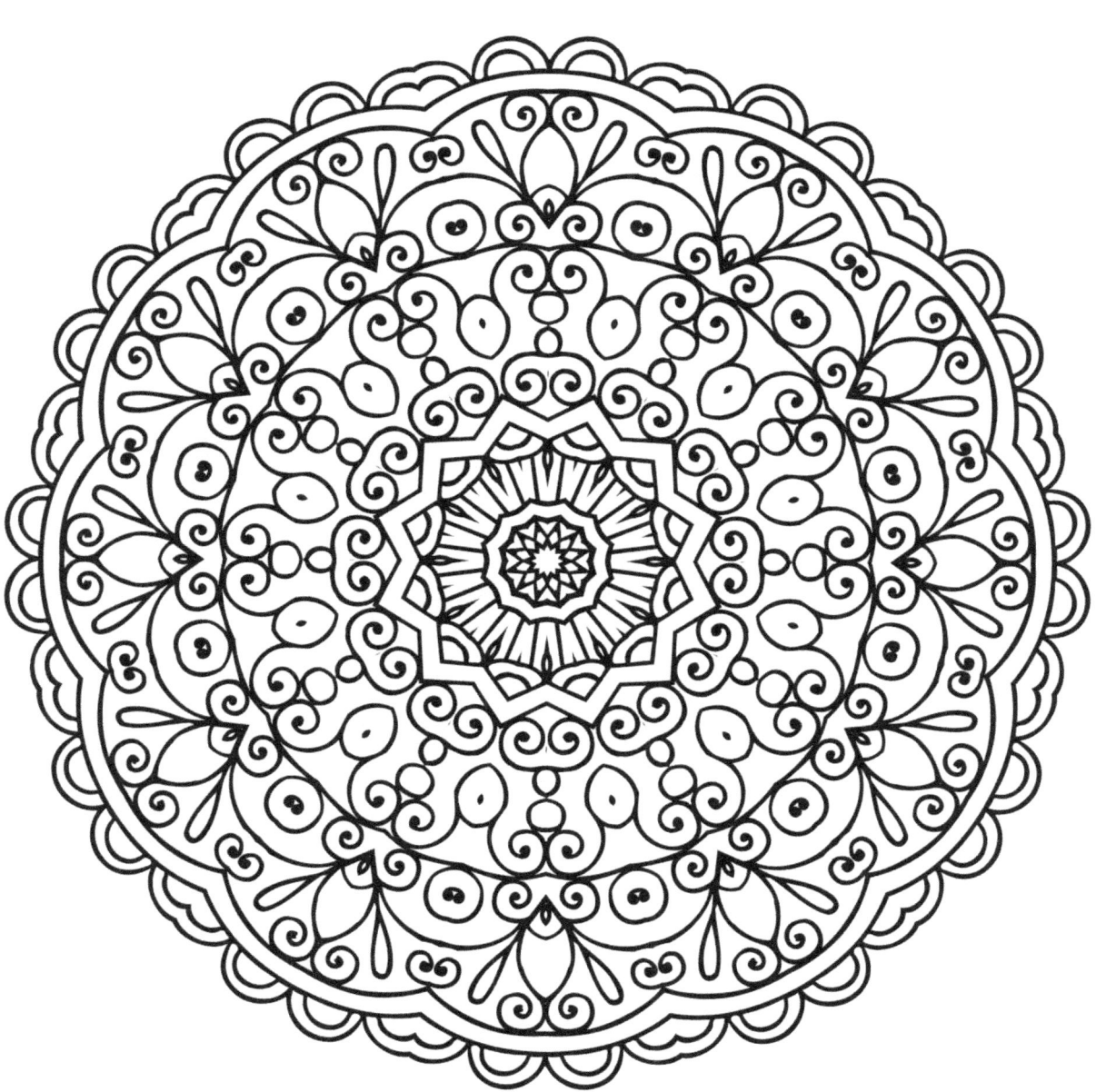

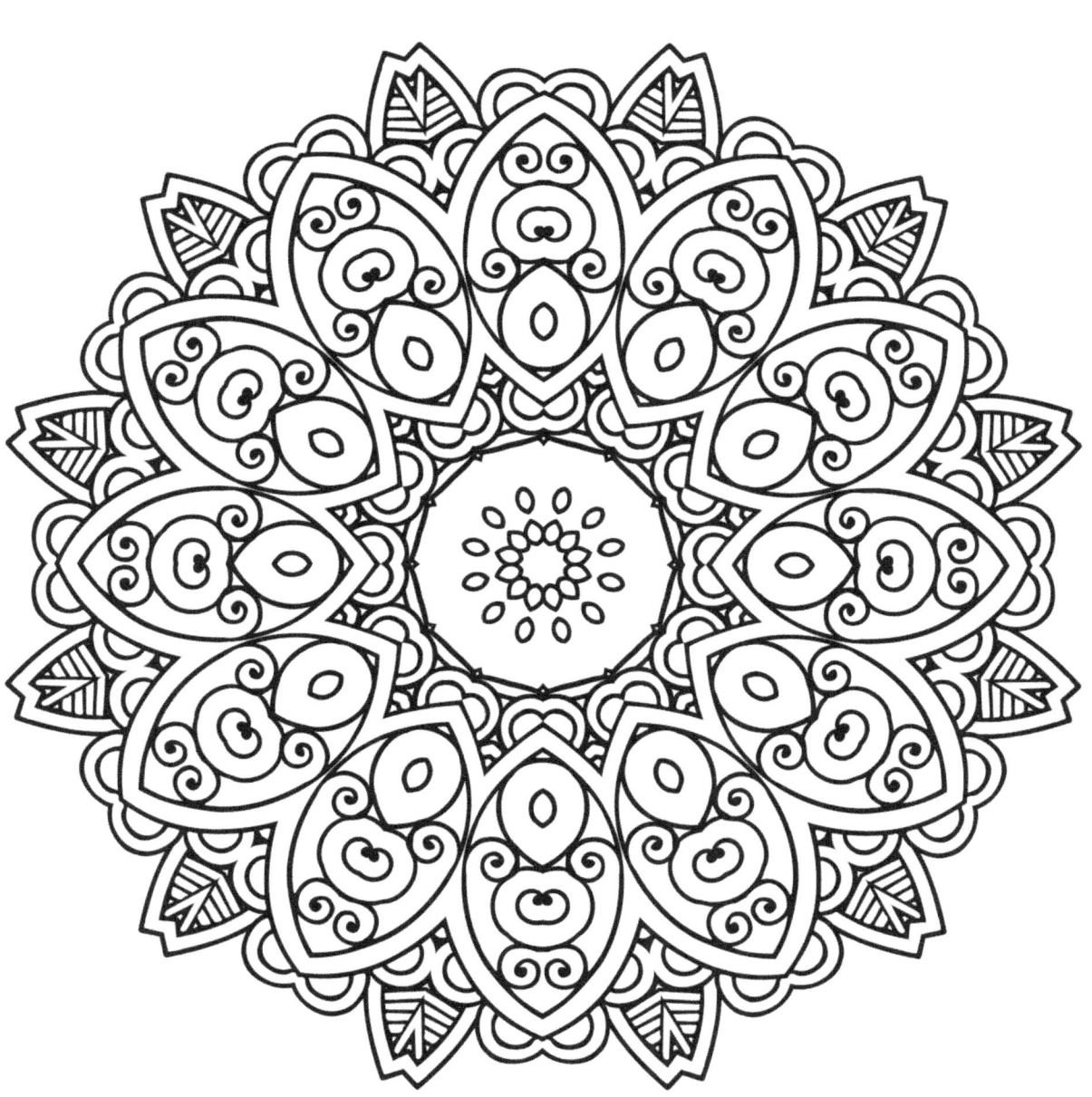

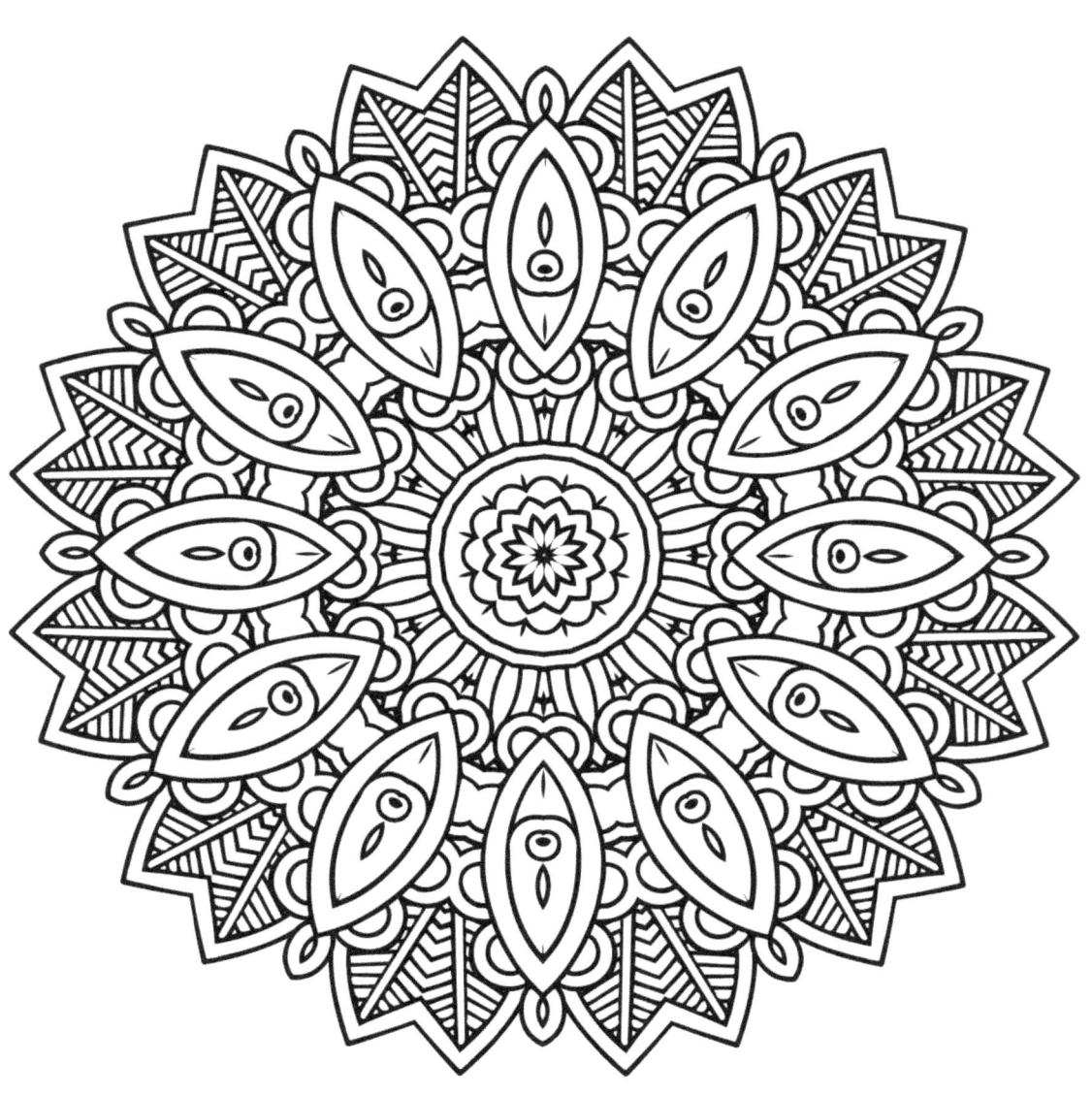

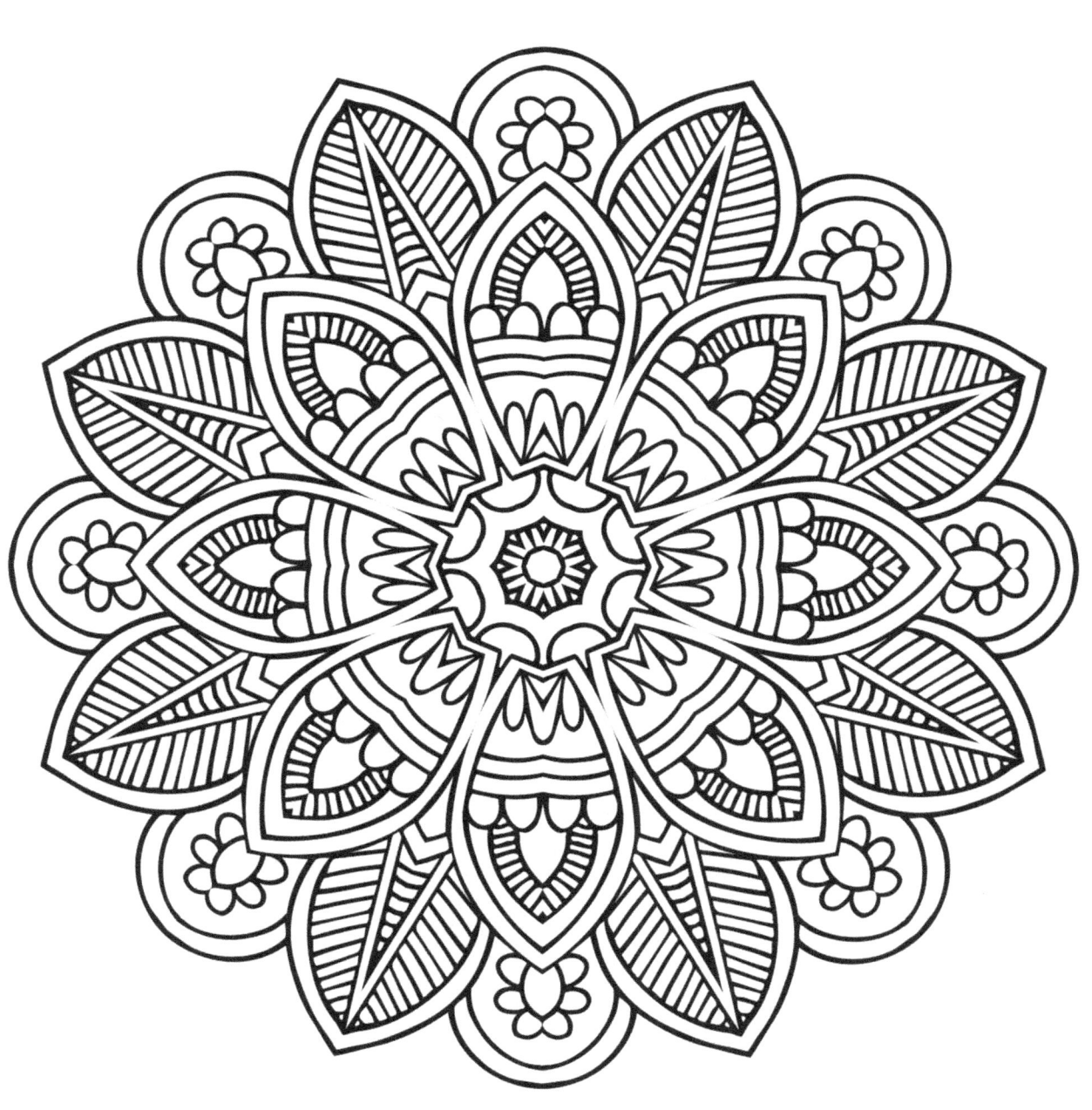

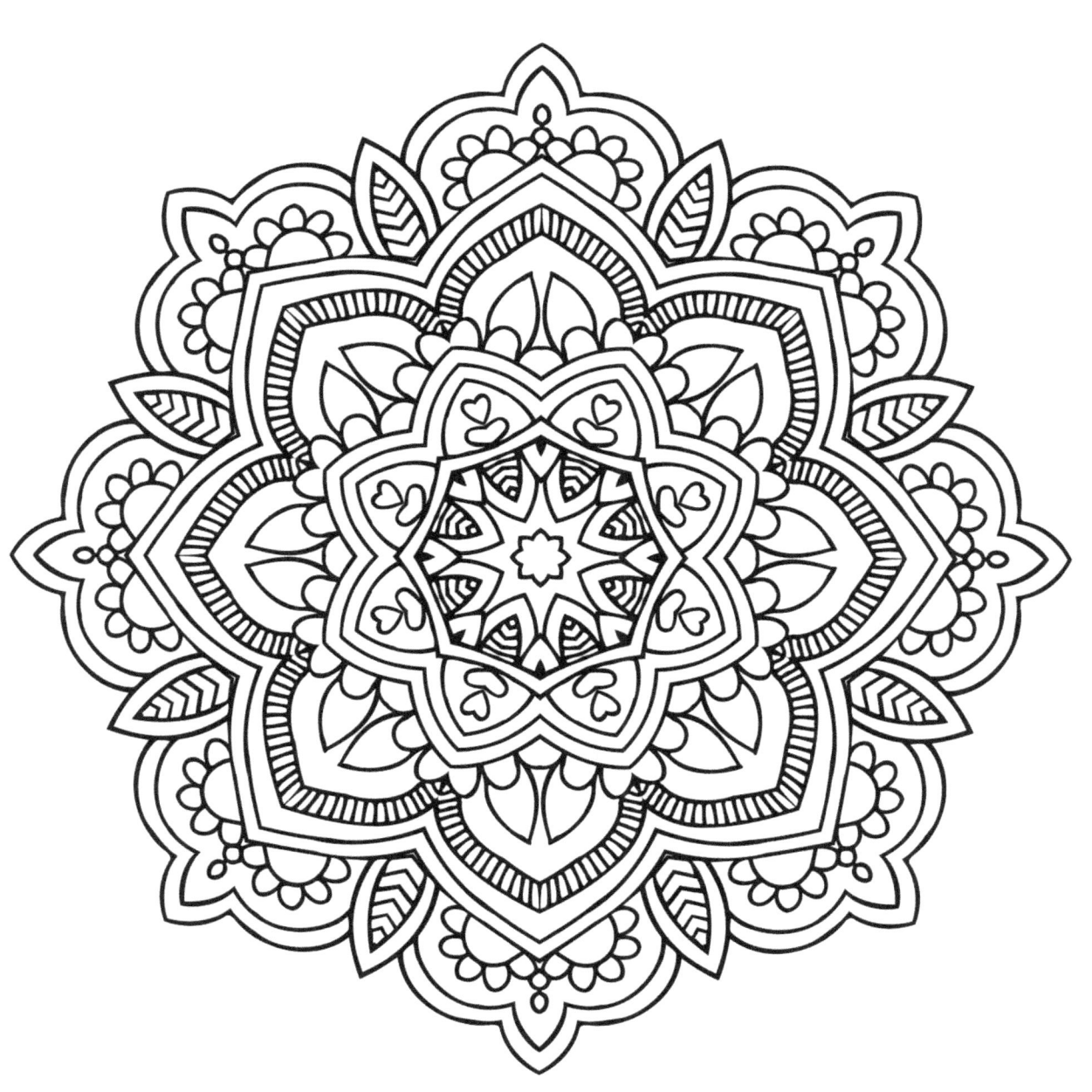

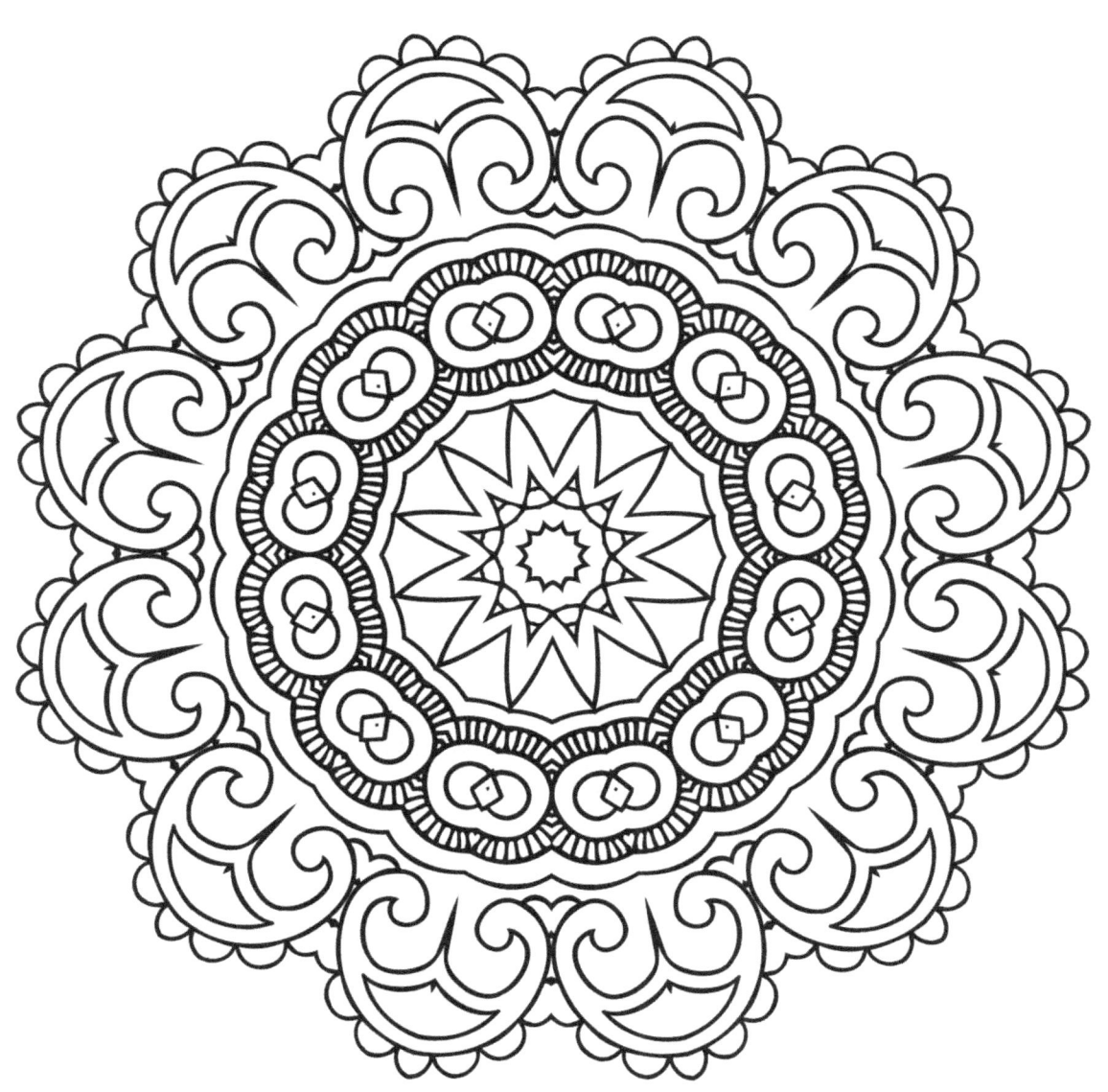

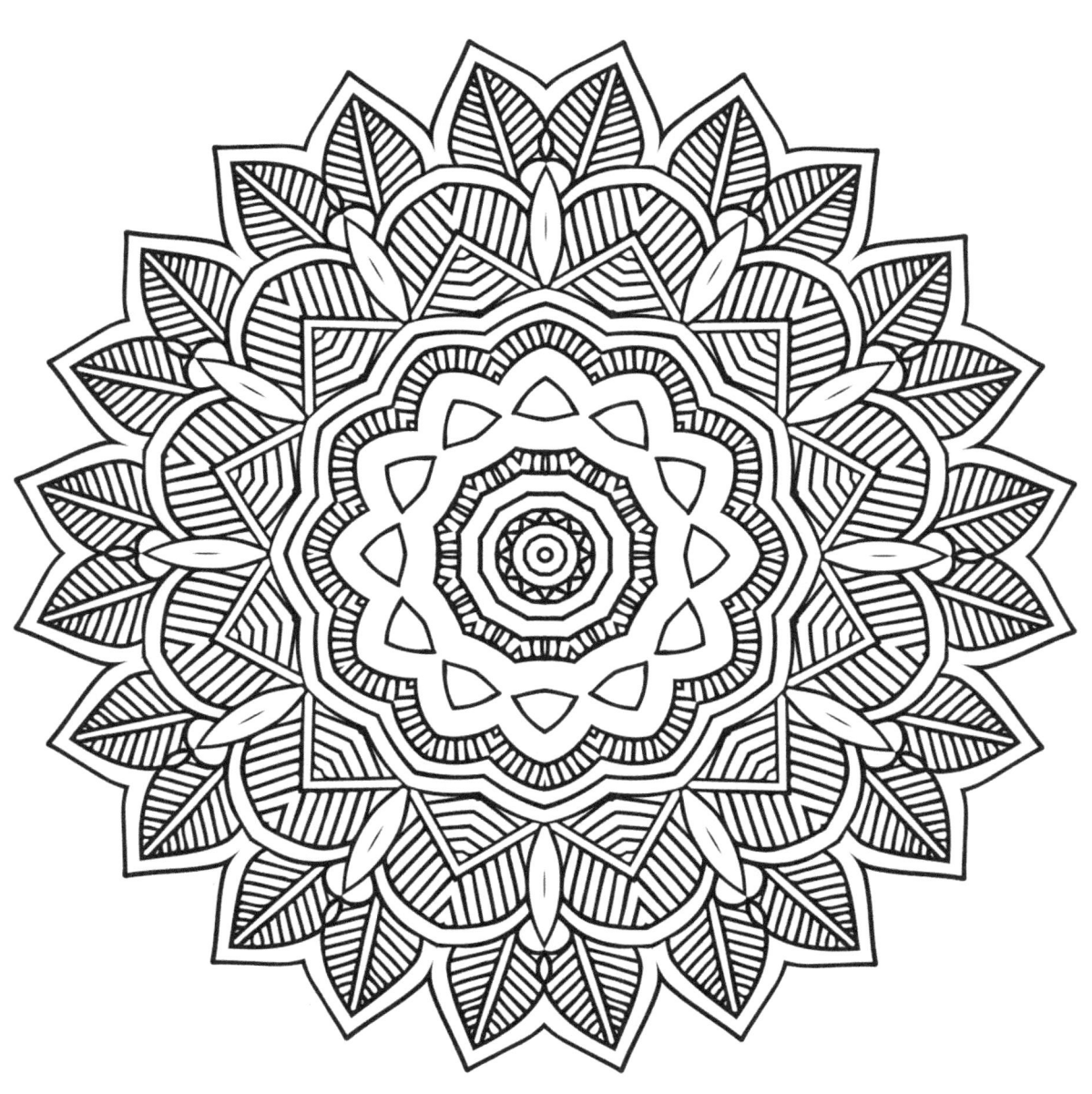

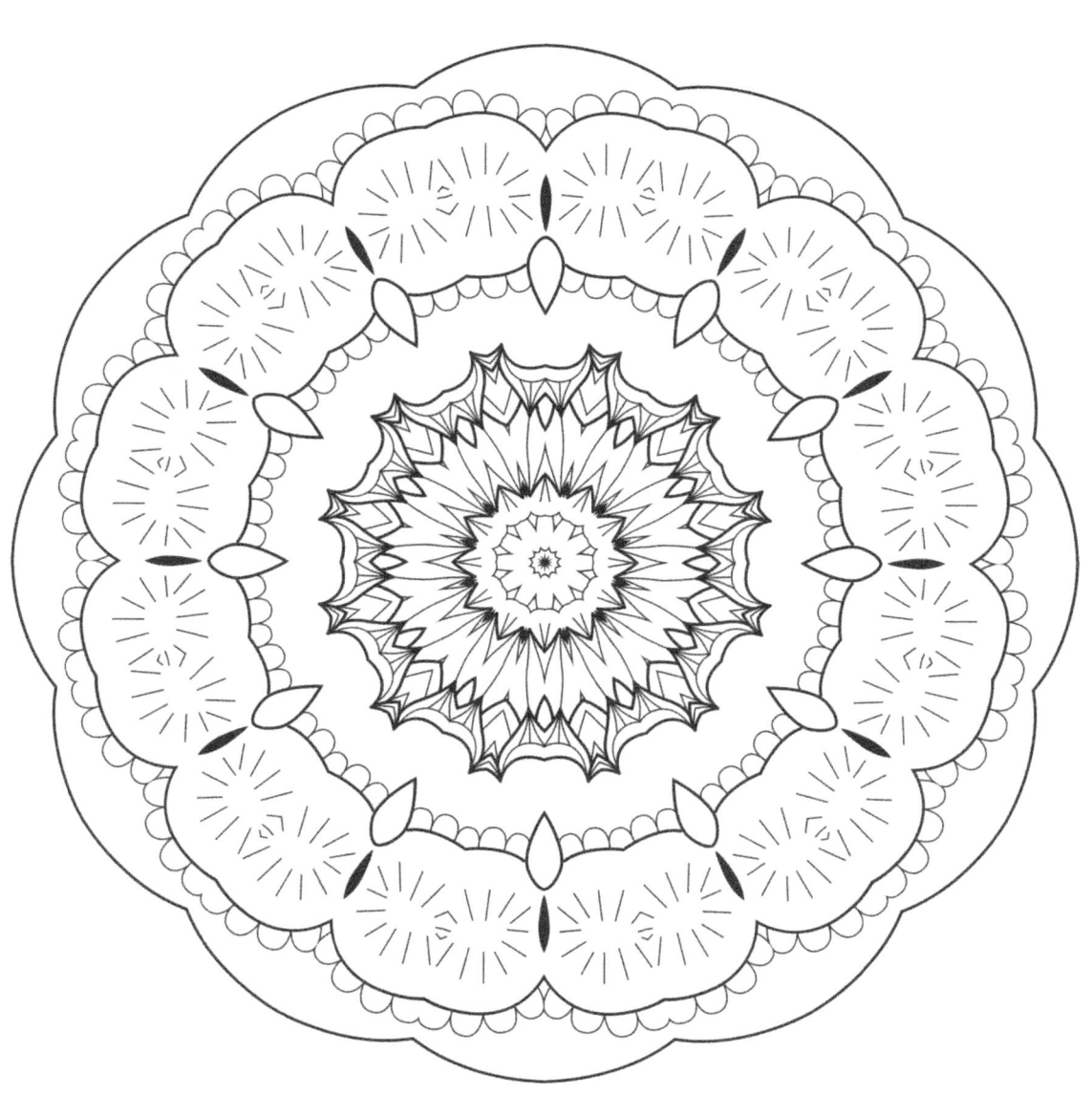

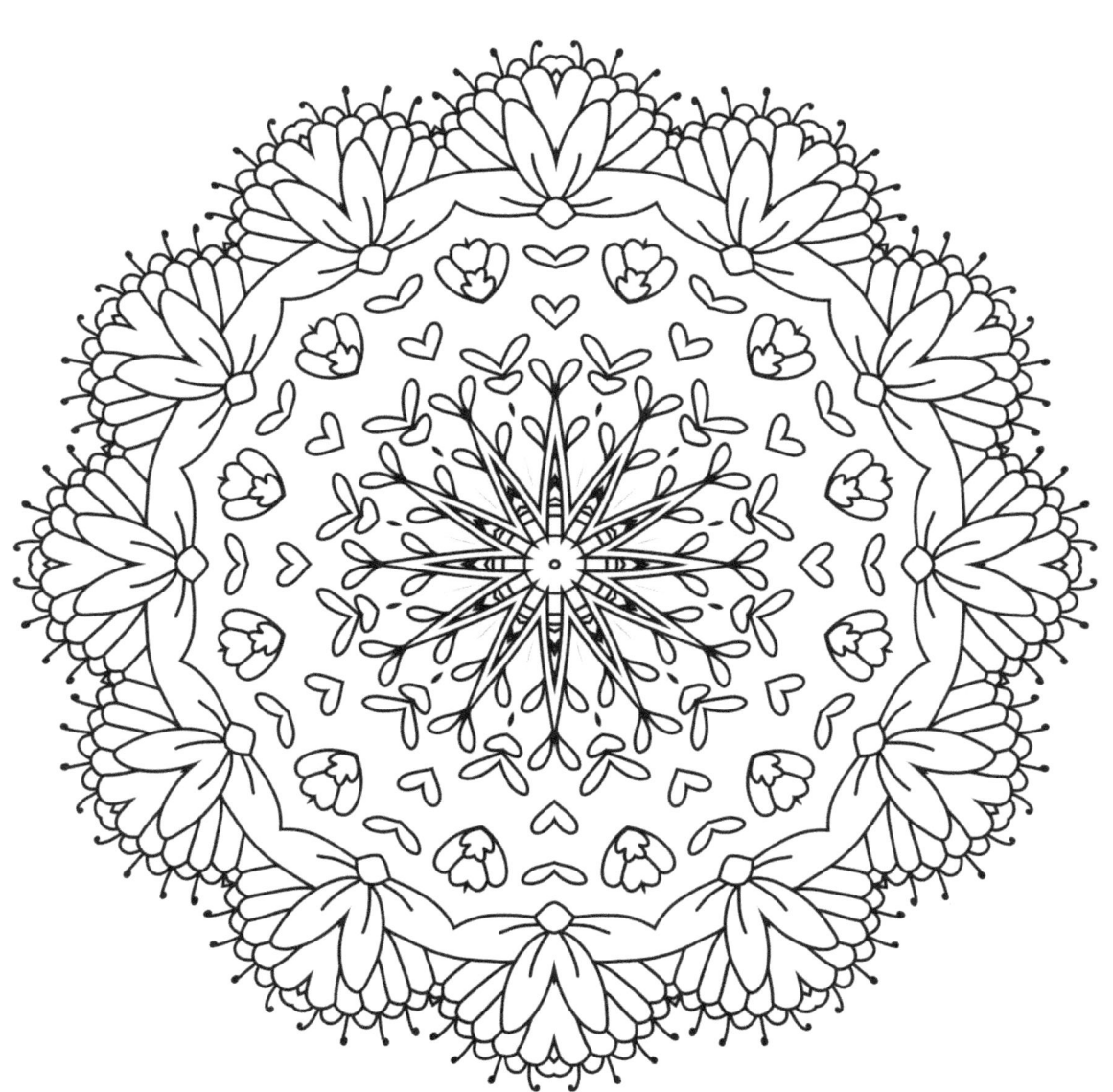

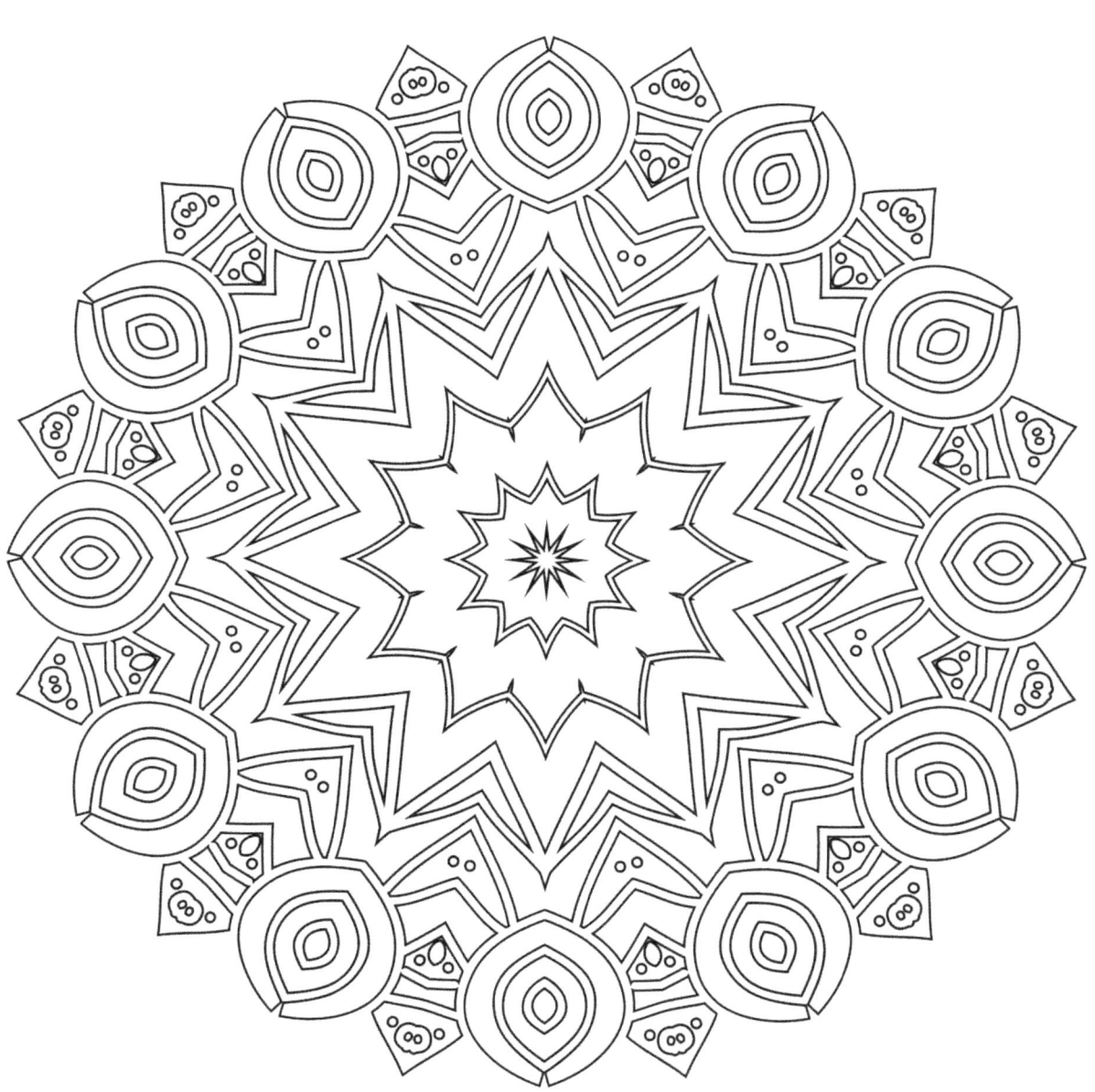

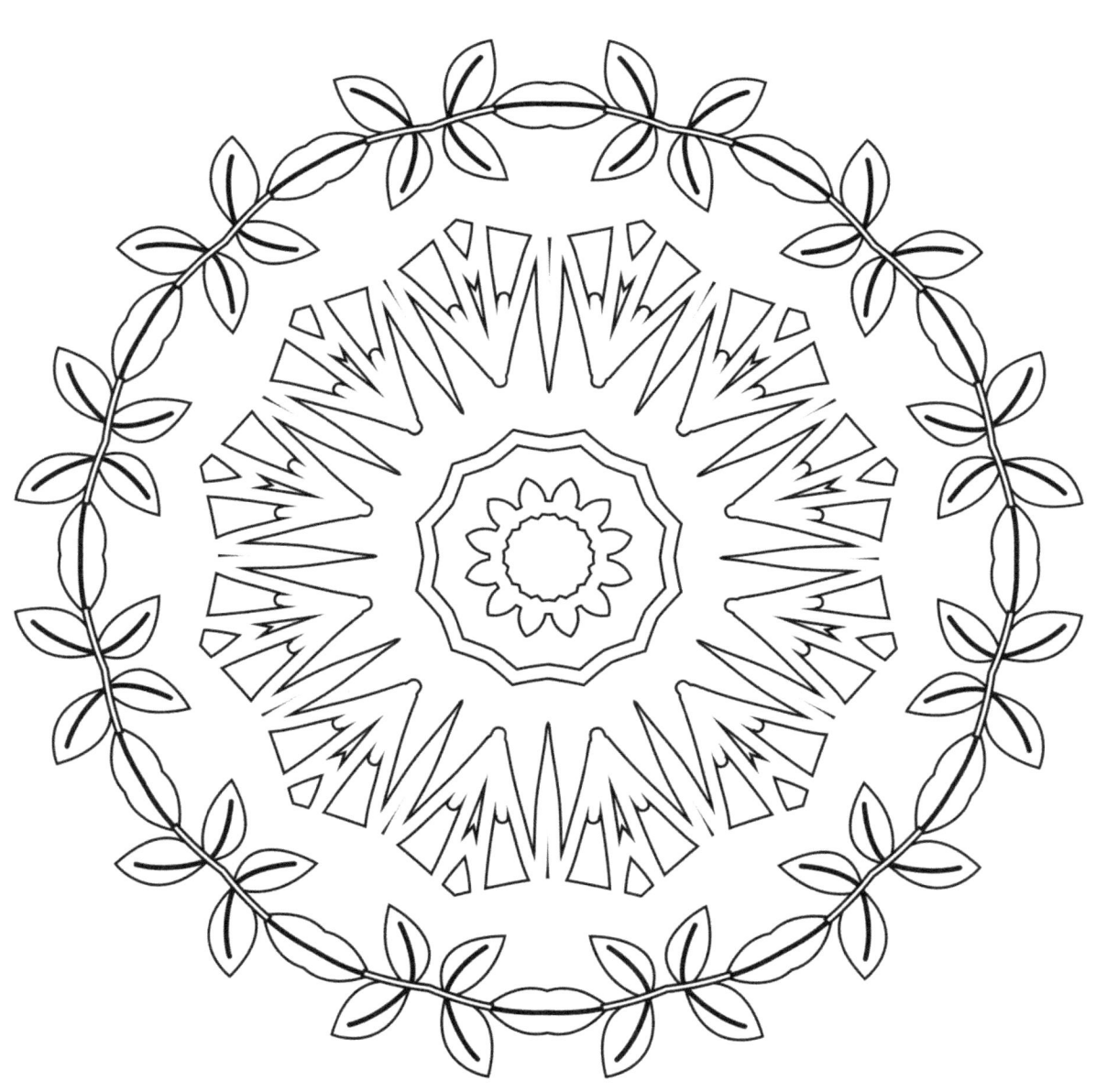

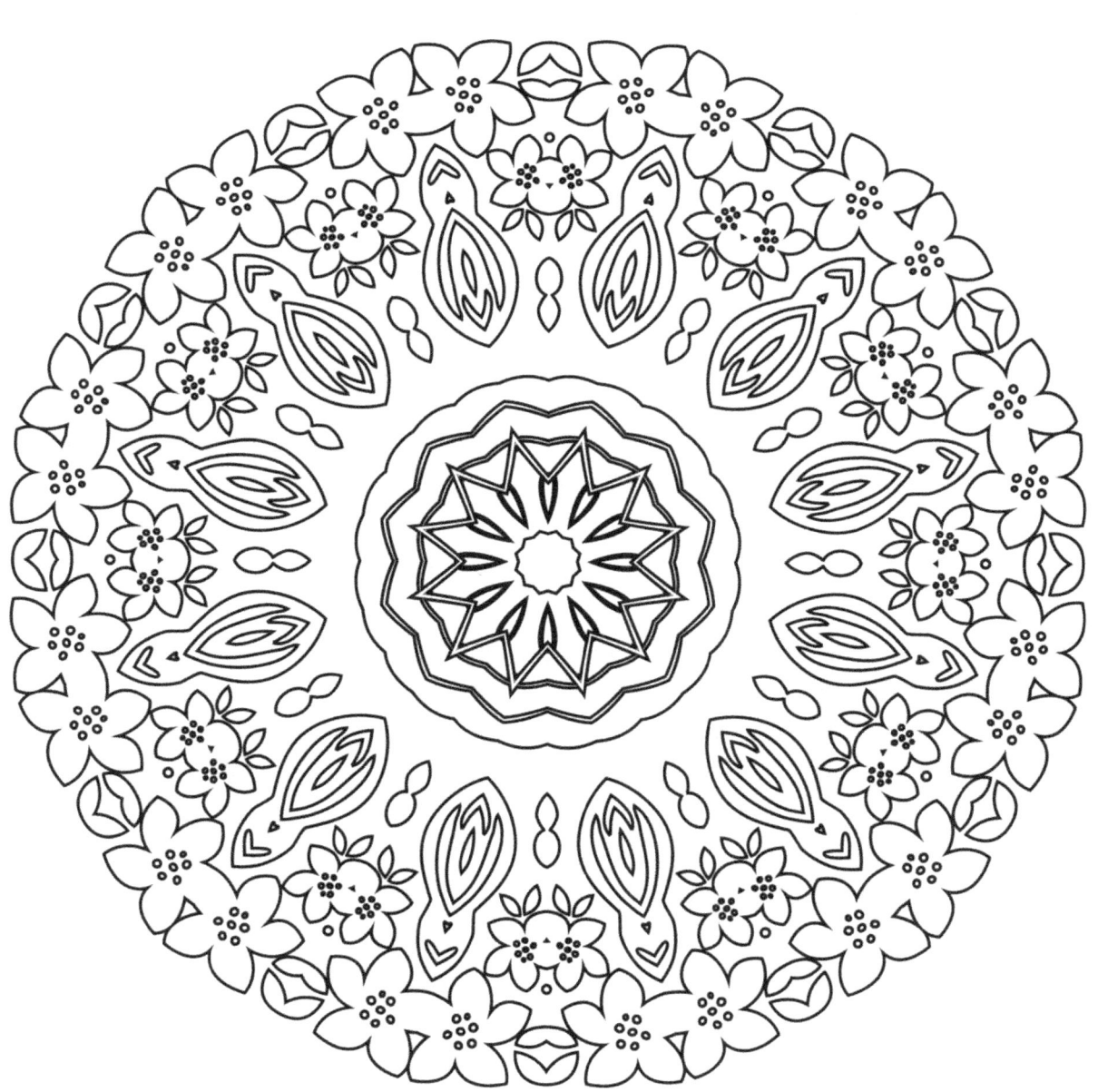

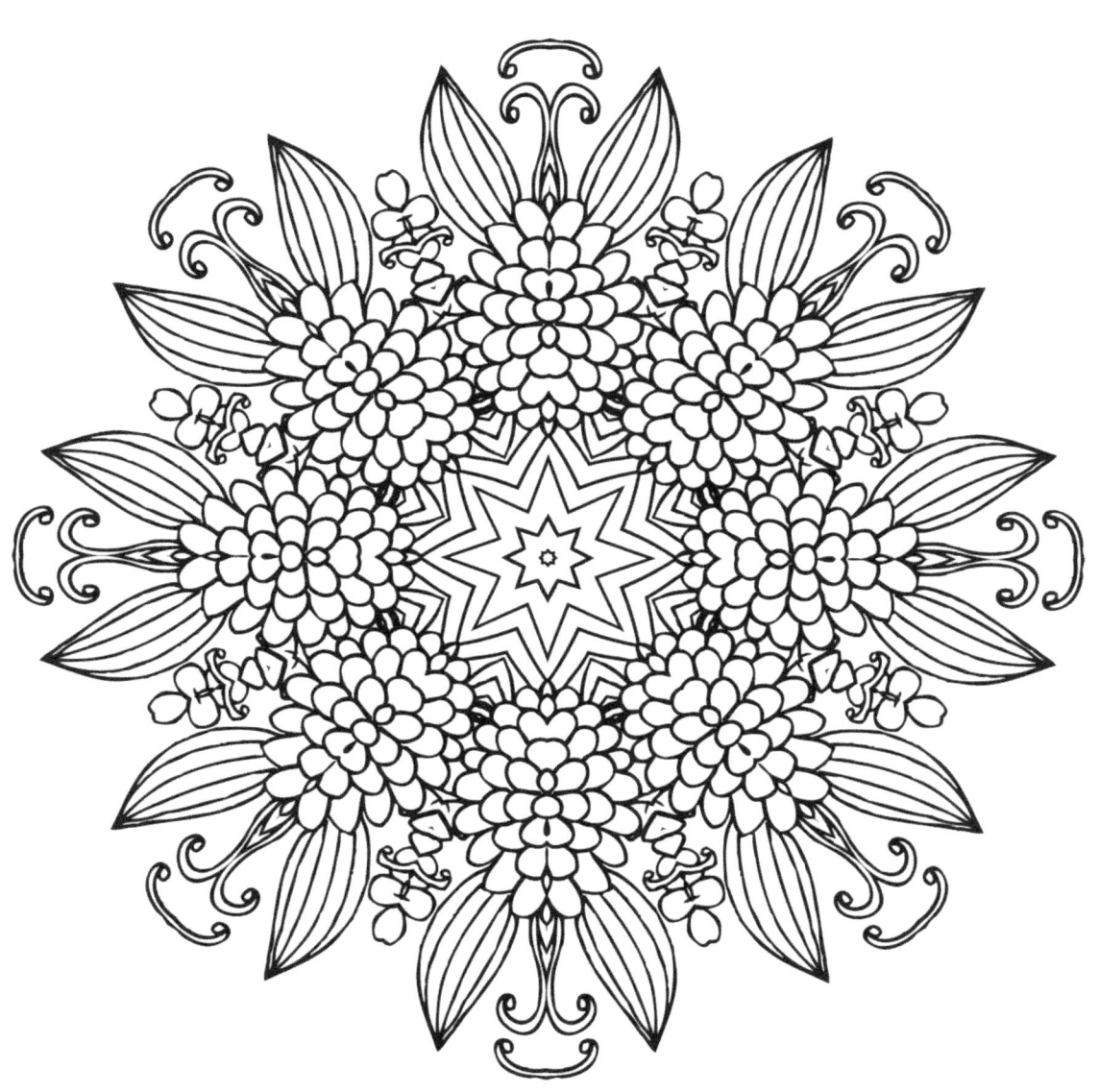

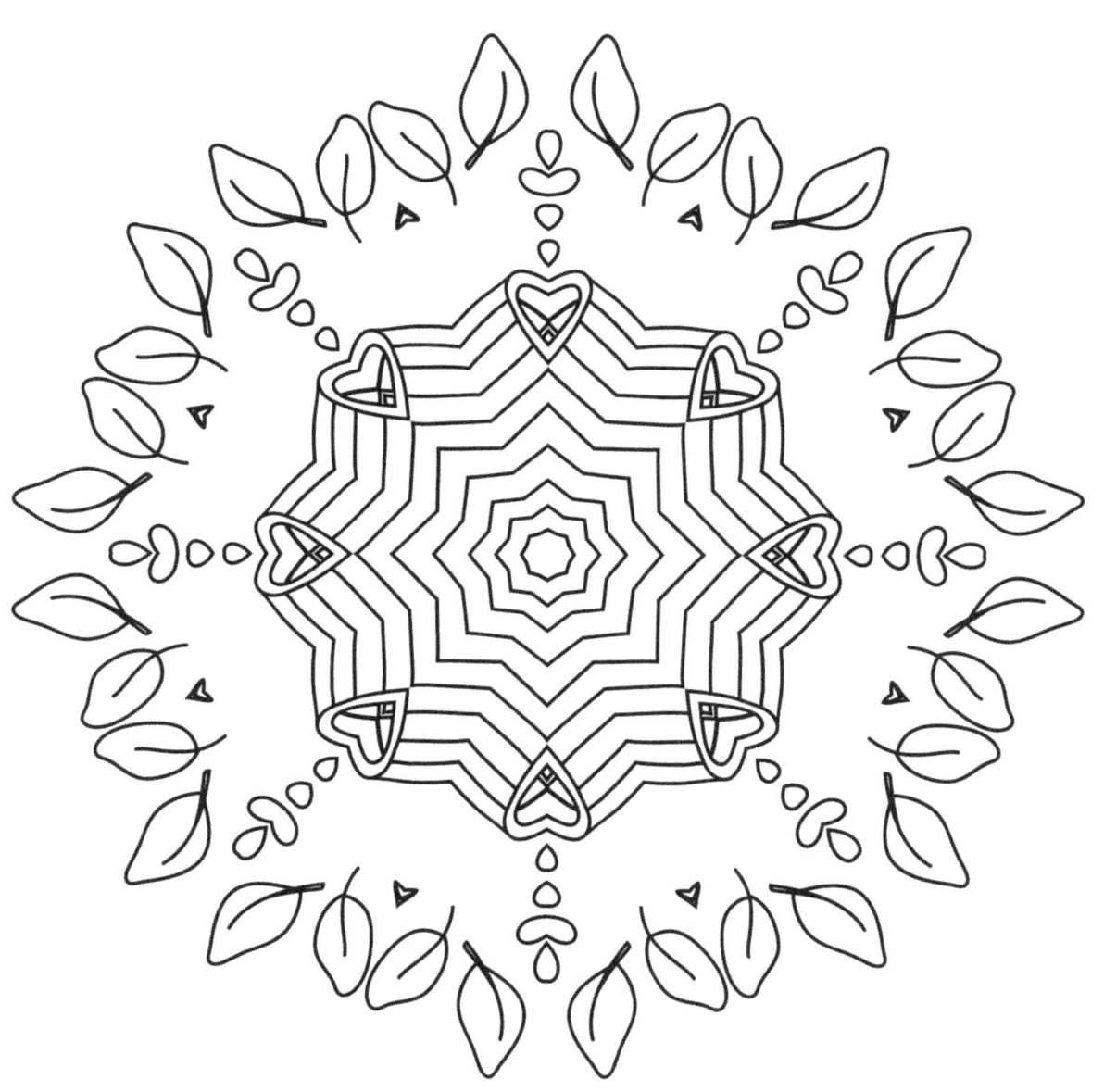

www.ingramcontent.com/pod-product-compliance
Lightning Source LLC
Chambersburg PA
CBHW080554190526
45169CB00007B/2766